José Leôncio de A. Silva
Sergio N. Duarte

Performance of washing equations for recovering salinized soil

AF666068

José Leôncio de A. Silva
Sergio N. Duarte

Performance of washing equations for recovering salinized soil

Salinity in agriculture irrigated by excess fertilizers in a protected environment

ScienciaScripts

Imprint

Any brand names and product names mentioned in this book are subject to trademark, brand or patent protection and are trademarks or registered trademarks of their respective holders. The use of brand names, product names, common names, trade names, product descriptions etc. even without a particular marking in this work is in no way to be construed to mean that such names may be regarded as unrestricted in respect of trademark and brand protection legislation and could thus be used by anyone.

Cover image: www.ingimage.com

This book is a translation from the original published under ISBN 978-3-639-83926-5.

Publisher:
Sciencia Scripts
is a trademark of
Dodo Books Indian Ocean Ltd. and OmniScriptum S.R.L publishing group

120 High Road, East Finchley, London, N2 9ED, United Kingdom
Str. Armeneasca 28/1, office 1, Chisinau MD-2012, Republic of Moldova, Europe
Printed at: see last page
ISBN: 978-620-8-29196-9

Copyright © José Leôncio de A. Silva, Sergio N. Duarte
Copyright © 2024 Dodo Books Indian Ocean Ltd. and OmniScriptum S.R.L publishing group

SUMMARY

1 INTRODUCTION

In regions where irrigation is intensive, with the aim of increasing crop production, fertigation is widely used as an alternative to fertilization, as it presents a lower degree of risk. The technique combines the supply of the crop's water demand with the application of nutrients via irrigation water, leaving the nutrients in soluble form and therefore more available to the crop. However, excessive fertilizer application via fertigation can increase the risk of soil salinization, especially if fertilizers are applied indiscriminately.

The increase in soil salinity in protected environments is generally related to the excessive use of fertilizers applied via irrigation water. Therefore, it is necessary to take preventive and corrective measures, since the intensive use of fertilizers, the lack of rainfall and the continuous evaporation of water from the soil can increase the salt content in the soil solution, resulting in serious damage to irrigated agriculture and especially to the environment.

Under these growing conditions, when the technology of increasing fertilizers via irrigation water is used without any form of management or technical knowledge, fertigation becomes a problematic contributor, due to the increase in areas with high saline concentrations in the soil, which can lead to a serious problem for the crop.

Although many producers use good quality irrigation water with low salinity, the application of fertilizers in a protected environment is generally done via irrigation water, which makes it saline, and the level of salinity of the irrigation water is approximately proportional to the amount of fertilizer added.

There are several negative effects of salinity in the soil; these are directly related to plant growth and yield and, in extreme cases, can cause total crop loss. In semi-arid regions, in order to minimize these problems, producers have used the practice of routinely leaching salts into deeper soil layers. However, when the leaching fractions used are not sufficient to prevent an increase in the salt content of the soil during a crop's cycle, it is necessary to reduce the salinity of the soil to a level tolerated by the crop to be planted, by means of reclamation washing (RHOADES et al., 1992).

According to Hoffman et al. (1992), when the water is not too saline, so that during the crop cycle the soil salinity does not exceed the threshold salinity, it is not necessary to carry out constant leaching; in this case, only a maintenance wash can be carried out at the end of the crop cycle.

Due to these problems, the FAO warns that approximately 50% of the world's 250 million hectares of irrigated land already have salinization and soil saturation problems, and that around 10 million hectares are abandoned every year because of these problems.

Rainfall in open fields is not always enough to wash salts out of the soil. Therefore, when soil salinity exceeds the crop's tolerance limit, cultivation should be stopped temporarily and the process of removing salts accelerated through the use of recovery practices.

The volume of water required for the reclamation washing of a soil profile is calculated depending on the initial salinity of the soil, the desired final level, the type of soil and the depth of soil to be reclaimed, the method of application of the irrigation water and the concentration of salts in the washing water (DIAS, 2003). Although a lot of research has already been carried out using mathematical models to simulate the recovery of saline soils, the amount of water used in washing required to recover these soils is generally calculated using empirical relationships derived from field experiments (RHOADES & LOVERDAY, 1990).

Irrigation with intermittent water application has great potential for increasing the efficiency of wash water application, due to the increased time needed for solutes from the micropores to diffuse into the macropores. This generally allows a greater decrease in EC to be achieved using less water (KEMPER et al., 1988).

In view of the technologies that can be employed, and given the information on salinity control and soil management, and considering that there are currently several models and formulas for calculating the washing slurry, resulting in a wide variation in the amount of water recommended, the aim is to evaluate the performance of these equations for recovering soils affected by salts, in an attempt to identify those that provide the most accurate management of fertilizer salts, carried out in a sustainable manner, with the least possible risk to the environment.

2 LITERATURE REVIEW

2.1 Origin of salts in the soil

Salts originate from the weathering of primary minerals found in soils and rocks. These salts are transported by rainwater and stored in the soil, accumulating as the water is evaporated or consumed by crops, giving rise to the salinization process (DAKER, 1988). The salts released during the rock weathering process, depending on the geomorphology of the region, can be carried to lower horizons by percolation or be carried to distant places by surface runoff. Depending on the conditions of relief, water flow, etc., salts are deposited in sub-surface water and can, by capillarity, accumulate on the soil surface as the water is evaporated or consumed by the plant; the second phenomenon, i.e. surface runoff, is responsible for the deposition and accumulation of salts in rivers, seas, reservoirs and lagoons (AYERS & WESTCOT, 1991).

In regions with arid and semi-arid climates, because they have a water deficit for most of the year, partly due to the fact that the soils are shallow or have impermeable layers, water containing salts is subject to the processes of evaporation or evapotranspiration and can, over time, reach levels in the soil solution that compromise plant growth and development (GHEYI, 2000).

Although the main and direct source of all the salts present in the soil is the weathering of rocks, there are few examples where this source of salts has directly caused problems related to soil salinity. Commonly, soil salinization occurs due to the accumulation of salts transported by water from other places, as well as the invasion of tidal waters in coastal regions and concentration by evaporation in regions with low rainfall, when the water table is close to the surface. Soil salinization due to this phenomenon is called primary salinization (RIBEIRO et al., 2009).

Salinity problems have also been associated with the water used for irrigation, poor drainage, the presence of sub-surface water rich in soluble salts and the shallowness of the water table. In these cases, i.e. when the resulting salinization is due to anthropogenic action, it is known as secondary salinization. In addition, salinization can be caused by fertilizer being applied excessively and sparingly throughout the crop cycle, inducing the root system to undergo osmotic stress (AYERS & WESTCOT, 1991).

In some cases, secondary salinization is responsible for irreparable losses in the productive capacity of soils, rendering them sterile in large tracts of cultivated land. Thus, water of dubious quality (with risks of salinity and sodicity), fertilizers with a high saline content (potassium chloride, sodium nitrate and ammonium nitrate, among others), inefficient drainage, etc., can accelerate the process of secondary salinization (OLIVEIRA, 1997).

Water is an extremely important resource for sustaining life, the environment and a range of activities that drive a country's economy, especially agriculture (MACHADO et al., 2003). According to Becker (2005), water is considered to be the "blue gold of the 21st century", due to its scarcity and growing use around the world, especially in semi-arid countries, which use irrigation for agricultural crops. Furthermore, there are predictions that the dispute over water could lead to conflicts. In this context, there is a need to develop more efficient ways of capturing, storing and using water, in order to avoid a shortage for future generations.

Irrigation with intermittent water application comprises various forms of incomplete irrigation, called cycles, in which the advance phase continues after the water has been cut off in the application section and often during the depletion and recession phases (WALLENDER AND RAYEJ, 1985). According to Bishop et al. (1981), the cyclic time can vary from a few seconds to hours, but should be between 10 and 60 min. The cyclic rate is defined as the ratio between the application time and the cyclic time.

Coolidge et al. (1982) highlighted two practical advantages of intermittent application: varying the time of application makes it possible to design the system, predicting a reduction in the average flow rate after the advance phase has been completed, thus reducing losses through surface runoff; it is observed that, for the same volume of water applied, when compared to continuous application, the efficiency of application increases. Thus, intermittent surface application is more commonly used in places with an unlimited water supply and in systematized areas, which facilitates irrigation management (STONE, 2005).

2.2 Salinity in irrigated agriculture

The use of irrigation has made a significant contribution to increasing agricultural productivity and incorporating into the production system areas whose potential for agriculture is limited due to their rainfall patterns. On the other hand, irrigation has caused some problems for the environment. These include the inappropriate use of saline and/or sodic water, resulting in a loss of the soil's productive capacity. Water salinity can cause changes in the soil's physico-chemical properties (RHOADES et al., 1992, cited by LIMA, 1998).

In general, irrigated agriculture depends on both the quantity and quality of water. Until the beginning of the 20th century, due to the availability of good quality water that was easy to use, there was no concern about the preservation and rational use of water. From that time onwards, with the increase in the consumption of quality water, there was a reduction in its availability, which consequently led to the need to use lower quality water (AYERS & WESTCOT, 1991). Among the characteristics that determine the quality of water for irrigation, the concentration of soluble salts or salinity is one of the most limiting factors for the development of some crops (BERNARDO, 1996).

The main problems caused by soil salinization are a reduction in the osmotic potential of the soil solution, reducing its availability to plants, and an increased risk of toxicity of certain ions to plants (BERNARDO, 1996).

The accumulation of salts in the rhizosphere damages the growth and development of crops, causing a decrease in productivity and, in more severe cases, can lead to a collapse in agricultural production. The salts are transported by the irrigation water and deposited in the soil, where they accumulate as the water evaporates or is consumed by the crops. The damage results from a decrease in the osmotic potential of the soil solution or from the toxic effects of specific ions or due to changes in the physical and chemical conditions of the soil (LIMA, 1998).

Normal soils can become unproductive if they receive too many soluble salts due to improper irrigation with saline water. Even with good control of irrigation water quality, which is rarely done in practice, there is a continuous residue of salts in the soil (SOUZA, 1995).

Secondary salinization is caused by human action and is mainly manifested as a result of irrigation, in areas where drainage control is not carried out or where drainage is carried out inefficiently. In the world, it is estimated that 19.5% of irrigated land (45 million hectares) and 2.1% of non-irrigated land (32 million hectares) are affected by salts. Soil salinization is particularly evident in arid and semi-arid regions, affecting around 25% of irrigated areas. Currently, it is estimated that around 1.5 million hectares of arable land are lost each year due to the accumulation of salts. Thus, salinization has been identified as the main process of soil degradation in the world (FAO, 2006).

The process of soil, groundwater and surface water salinization is one of the most important problems of environmental degradation, and its harmful effects are most pronounced in arid and semi-arid regions. This problem has been growing rapidly in various parts of the world, causing major problems for the productivity of agricultural crops. On a global scale, there is an area of approximately 400 million hectares of land used for agriculture, the production of which is being severely restricted by salinity (BOT et al., 2000).

In Brazil, although information on saline areas is not well defined, it is estimated that 20 to 25% of irrigated areas face salinization problems (FAO, 2006). This problem is most prevalent in the Northeast region, which has an area of 155 million hectares, 52% of which is in a semi-arid climate. Considering that the practice of irrigation is the only way to guarantee safe agricultural production in this climate and that the implementation of irrigation projects has increased in recent decades, the development of research that will enable better soil and water management is of great importance for agricultural production, given the future projections of an increase in population and demand for food (HOLANDA et al., 2010).

In the semi-arid Northeast, the highest incidences of areas affected by secondary salinization are concentrated in the most intensively cultivated lands, with the use of irrigation, in the so-called irrigated perimeters (OLIVEIRA, 1997). The consequence of soil salinization is a reduction in crop yields, making it necessary to carry out reclamation washing and, in some cases, the addition of chemical conditioners, such as agricultural gypsum, which leads to additional costs.

The practice of irrigation in arid and semi-arid regions is indispensable due to the occurrence of water deficit for crops in the dry season (OLIVEIRA, 1997). However, in these regions it is common to find water sources with high levels of salts which, before being used for irrigation, should be evaluated, especially in terms of salinity, sodicity and ion toxicity. The quality of irrigation water can vary significantly according to the type and quantity of salts dissolved in it. In order to foresee future problems regarding the quality of water for irrigation, it is important to consider the potential for water to create conditions in the soil that restrict its use and to assess the need to expand special management techniques in order to maintain good yields (AYERS; WESTCOT, 1991).

Efforts have been made to control salinity through technical means; however, the recovery of a soil rendered unusable by excess salts or sodium, or in the process of being rendered unusable, generally involves large investments and can only be economically viable under certain circumstances (PRISCO, 1987). In order to improve the soil's physical properties, various combinations of plowing practices, the use of correctives and cultural practices can be used (GHEYI, 2000).

2.3 Recovering soils affected by salinity

Irrigation water with high levels of salts can contribute directly to increasing the saline concentration of the soil, and together with excessive evaporation, can lead to the accumulation of soluble salts on the surface and/or sub-surface of the soil (BERNARDO, 2005, PEREIRA, 1998). To remove soluble salts, washing is the most effective method for recovering saline soils. The technique recommends applying enough water to dissolve and transport the soluble salts to the drainage system. The depth of reclamation depends on the type of crop being cultivated. For a crop with a superficial root system, the depth may be 60 cm; however, if the crop to be exploited has a deep root system, the soil should be corrected to a depth of around 2.00 m (RICHARDS, 1954; RHOADES,1974; RIBEIRO et al., 2003).

In the recovery of saline soils there are several alternatives that can be used, separately or in combination (AYERS & WESTCOT, 1991).

Proper management of salt-affected soils is essential for efficient and sustainable irrigated agriculture. It includes the recovery of salt-affected soils, usually caused by inadequate irrigation, and the maintenance or prevention of unaffected irrigated soils and recovered soils. It involves the application of existing theories and knowledge in soil physics and chemistry, as well as in the irrigation-salinity and production-salinity relationships (SANTOS & HERNADEZ, 1997).

Salt washing can be carried out for two purposes: 1) to reduce the initial high salinity of the soil to levels tolerated by the crops, known as reclamation washing; 2) to prevent the resalinization of reclaimed soils or to prevent the salinization of unaffected irrigated soils, which is commonly referred to as maintenance leaching (PIZARRO, 1978). This practice is necessary in the recovery of both saline and sodic soils (SANTOS & HERNADEZ, 1997).

Reclamation washing usually requires large quantities of water and, when soil salinity is very high ($EC_{es} > 10\text{-}15$ dS m^{-1}), it is carried out without agricultural use of the soil for a certain period of time (from weeks to months). When soil salinity is not too high (EC $<$ 1015 dS m^{-1}), reclamation washing can be done with extra water, applied to irrigate tolerant crops (AYERS & WESTCOT, 1991). The amount of water to be used, as well as the recovery time for saline soils, depends on the initial salinity of the soil, the quality of the irrigation water, the depth of the soil to be recovered, the irrigation techniques used, among others.

Thus, more appropriate management practices to control soil salinity in the long term are essential in a program for cultivating soils affected by salts, especially with a view to sustainability, as well as being alternatives for soil use and recovery. Among the techniques for recovering saline-sodic soils, the application of chemical correctives and subsequent washing of the soil are widely used, as they act directly to correct the problems of these soils in relation to plants (QADIR et al., 1996, MONTENEGRO & MONTENEGRO, 2004).

Pizarro (1996), quoted by Tavares Filho (2010), mentions that salt washing can be done in two ways to correct saline soils:

1. Continuous washing - covering the surface of the soil with a 10 cm high sheet of water. Here, the salts are removed more quickly and the area can be farmed again sooner. This method is recommended for soils with good permeability, a deep water table and a high evaporation rate; and

2. Intermittent washing - suitable for soils with low drainage capacity, a high water table and low salinity groundwater. It should be applied during periods of low evaporation.

Although physical-chemical models have been developed to simulate the recovery of saline soils, estimates of the amount of washing water needed to recover these soils are generally based on empirical relationships derived from research and practical field experience (RHOADES AND LOVEDAY, 1990). According to Palâcios (1969), the most suitable relationship is that proposed by Volobuyev, as it presents results that are closer to those obtained in the field and in the laboratory. Some of the equations found in the literature are presented below, such as Volobuyev's eq. (01), which has the following expression:

$$L = \alpha \log \frac{CEi}{CEf} \tag{01}$$

in which:

L - Amount of water needed to wash one meter deep of soil, cm;

α - Coefficient depending on chloride content and soil texture,

Table 1;

CEi - Initial electrical conductivity of the soil saturation extract, dS m^{-1} ; and

CEf - Desired final electrical conductivity in the extract, dS m^{-1} at 25°C.

Table 1 - α coefficient as a function of chloride content in % and soil texture

Textures	CHLORIDE CONTENT (%)			
	60 - 40	40 - 20	20 - 10	<10
Heavy	122	132	142	178
Average	92	102	112	148
Light	62	72	82	118

(Adapted from CORDEIRO, 2001)

Although equation (01) has been widely used, Cordeiro (2001) reports that this formula has several limitations, such as the fact that the "α" coefficient is only obtained as a function of the chloride content and soil texture, without considering the influence that other salts can have on the physical and chemical behavior of the soils to be washed. As an alternative, the author presents eq. (02), developed from the previous one, but with the advantage that factors such as: the chemical quality of the washing water, the salinity conditions of the soil, the depth of washing, etc. are integrated into the "α" coefficient.

$$L = 9{,}0\ (p)\ 0{,}75\frac{(70 - CEr)}{(CEi - CEr)}.\ 0{,}3.\log\frac{CEi}{CEf} \tag{02}$$

in which:

L - Amount of water needed for washing, cm;

p - Depth of soil to be reclaimed, cm;

CEr - Electrical conductivity of the wash water, dS m ;$^{-1}$

CEi - Initial soil electrical conductivity, dS m^{-1} ; and

CEf - Desired final electrical conductivity of the soil, dS m .$^{-1}$

Jury et al. (1979) also established the relationship shown in the following equation

(03):

$$\frac{C}{C_0} \cdot \frac{D_L}{D_S . \theta} = 0{,}8$$

(03) in which:

C - Concentration of salts in the soil solution after recovery, mg L^;[1]

C_0 - Concentration of salts in the soil solution before recovery, mg L^{-1} ;

D_L - Amount of water to be applied to the soil to leach salts, cm;

DS - Depth of soil to be reclaimed, cm; and θ - Volumetric water content in the soil, cm^3 cm .$^{-3}$

Based on experimental field data, Hoffman (1980) proposed the use of a coefficient in the previous relationship, resulting in eq. (04), proposed for the recovery of saline soils:

$$\frac{C}{C_0} \cdot \frac{D_L}{D_S} = k \tag{04}$$

in which:

C - Concentration of salts in the soil solution after recovery, mg L ;$^{-1}$

C0 - Concentration of salts in the soil solution before recovery, mg L ;$^{-1}$

D_L - Washing water to be applied to the soil to leach salts, cm;

DS - Depth of soil to be reclaimed, cm; and

k - Constant that varies with soil type and water application method.

The value of "K", depending on the type of soil, can be 0.45 for organic soils, 0.3 for clay soils and 0.1 for sandy loam soils. The K coefficient varies from 0.3 to 0.1 when reclamation is carried out by flooding and the soil varies from clayey to sandy. When water is applied by sprinkling or dripping, a value of 0.1 is suggested, regardless of the type of soil. However, Blanco & Folegatti (2001), when evaluating different washing slopes and water application methods, concluded that drip application was more efficient in recovering the soil than was predicted by the adoption of k=0.1.

Reeve et al. (1957) also established the relationship shown in the following equation (05):

$$\frac{La}{Ls} = \frac{CEi}{5CEf} + 0,15 \qquad (05)$$

in which:

La - Height of water to be leached, cm;

Ls - Thickness of soil to be recovered, cm;

CEi - Concentration of salts in the soil solution before recovery, mg L; and CEf - Concentration of salts in the soil solution after recovery, mg L^{-1}

3 MATERIAL AND METHODS

3.1 Location and preparation of the experiment

The experiment was carried out in tubular lysimeters with the surface protected with plastic, located in an open field belonging to the Biosystems Engineering Department of the Escola Superior de Agricultura "Luiz de Queiroz" - ESALQ/USP, Piracicaba - SP, situated at the geographical coordinates of 22° 42' south latitude and 47° 37 west longitude, at an altitude of 540 m (Figure 1).

Figure 1 - Location of the Experimental Area in the Biosystems Engineering Department of the "Luiz de Queiroz" School of Agriculture

According to the Koppen classification, the region's climate is of the CWa type, i.e. tropical humid, with three drier months (from June to August), summer rains and winter droughts, with an average temperature of the hottest month above 22 °C and the coldest month below 18 °C.

The study consisted of three stages: 1) Salinization of the soils stored in the concrete pipes, which were waterproofed inside; 2) Recovery of the salinized soils; and 3) Measurement of the salt concentrations in the soils after the desalination procedure and comparison of these concentrations with those calculated by empirical formulas. The first stage was based on the use of artificial salinization curves already obtained for the soil in question (Yellow Red Latosol of the "Sertâozinho" Series) in percolation columns.

The second stage consisted of applying the percolation sheets to wash and recover the salinized soils that were stored in the pipes.

In the next step, empirical equations taken from technical and scientific publications were used to estimate the concentrations of the salts remaining in the soil solution after the application of the washing sheets. The concentration predicted by the equations was compared with the actual concentration measured in the soil of the pipes.

3.2 Construction of the salinization curve

The following methodology was used to construct the salinization curve in order to obtain electrical conductivity values as a function of salt concentration.

To find the relationship between the electrical conductivity of the solution (ECs) and the total salts to be applied, these were dissolved in the desired proportions of salts, initially taking as a reference the equation proposed by Richards (1954). This empirical formula is presented as eq. (06).

$$C = 640.CE_s \tag{06}$$

in which:

C - Fertilizer salt concentration, mg L ;$^{-1}$

ECs - Electrical conductivity of the solution, dS m .$^{-1}$

Based on eq. (06), different concentrations were prepared which corresponded to a salinity called "theoretical salinity", in order to determine the actual electrical conductivity of the solutions using a bench conductivity meter. A curve was then drawn relating the concentration of the fertilizer solution and the electrical conductivity of these solutions, using a dispersion diagram. Subsequently, 10 kg samples of the soil to be used in this experiment were collected, air-dried, sieved and placed in 8 L pots. The pots had a perforated base and a 2 cm layer of gravel, covered with a geotextile blanket (BIDIM OP-30), resulting in a 20 cm layer of soil.

The soil was collected at a depth of up to 0.50 m, in the considerable agricultural layer of a profile classified as Yellow Red Latosol (EMBRAPA, 2006), sandy phase, from the ESALQ campus and locally called the "Sertaozinho" Series

The humidity of the soil in the pots was raised from air-dry to maximum retention capacity by applying the diluted solutions studied beforehand, with the aim of obtaining 9 levels of electrical conductivity in the saturation extract (ECes: 1.0:2.0:3.0:4.0:5.0:6.0:7.0:8.0 and 9.0 dS m^{-1}), with 3 replications. The final fertilizer concentration values in the water (cf) applied to the soil were those needed to obtain the EC levels to be reached in the field stage of the experiment, and were estimated based on the levels used in the salinization curve, with correction due to the difference in humidity.

The proportion of nutrients in the solutions adopted for salinization was based on the Hogland nutrient solution described by Furlani (1999) (Table 2).

Table 2 - Concentration of macro and micronutrients in the nutrient solution

Macronutrients (mg L $)^{-1}$						Micronutrients (mg L $)^{-1}$					
N	P	K	Ca	Mg	SO	B	Cu	Fe	Mn	Mo	Zn
200	31	293	100	24	32	0,2	0,03	3,4	1,1	0,05	0,2

Hogland, according to Furlani (1999)

The volume of water applied was that required to raise the soil humidity in the pots to the maximum retention capacity, and the quantities of salts that were added were sufficient to achieve the desired EC values. The final concentration of the fertilizers used was estimated using the ratio shown in eq. (07).

$$CF = \frac{U_S}{U_{CC}}.CI \qquad (07)$$

in which:

CF - Final fertilizer concentration in the salinizing solution, mg L ;$^{-1}$

Us - Weight-based moisture in the saturated paste correction, g g -1;

ucc - Soil moisture based on weight at field capacity, g g^{-1} ; and CI - Fertilizer concentration, based on the salinization curve in the previous step, mg L .$^{-1}$

After the solutions were applied to the soils, they waited 2 days and then soil samples were taken from the pots in the 0 to 20 cm layer, after scraping off the topsoil. From these, the electrical conductivity of the actual saturation extract was determined, i.e. that which would correspond to the values found in the salinization curve, using the saturation paste methodology proposed by Richards (1954). The result of this methodology is shown in Figure 02.

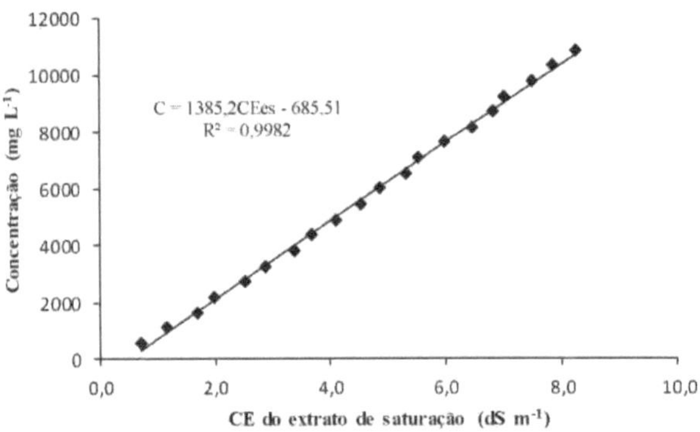

Figure 2 - Concentration of the solution to be applied to obtain the desired EC in the saturation paste of the soil used.

3.3 Implementation of treatments

The treatments were made up of a combination of two factors in a 2x5 factorial design, with two forms of soil recovery (R1=continuous and R2=intermittent) and five levels of soil salinity (S1=2.0: S2=4.0: S3=6.0: S4=8.0 and S5=10.0 dS m^{-1}). The experimental design adopted was a randomized block design with 6 replications.

In the experimental area, 60 waterproofed concrete containers, 1.2 m high and 0.5 m in diameter, totaling 0.20 m^2 of surface area and 240 liters of useful volume, were used from a previous study. The containers were positioned in 4 rows of 15 containers, so that each one represented an experimental plot (Figure 3).

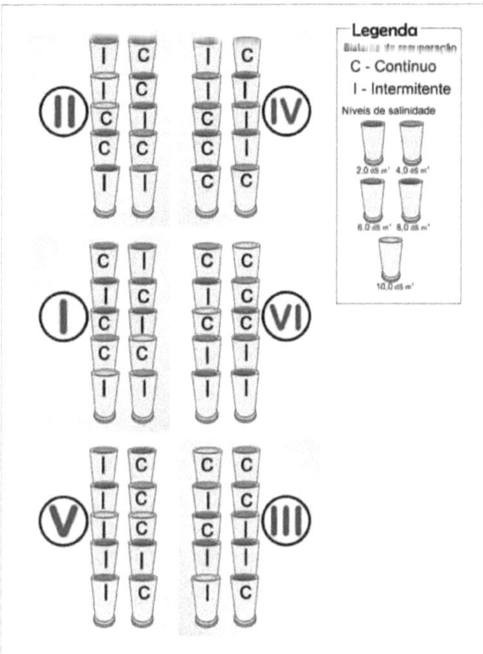

Figura 3 - Sketch of the experimental area

The drainage system for the pipes was made up of holes, located near the base of each pipe, where 19 mm diameter PVC pipes were inserted, sealed with epoxy putty and gutter sealant. The pipes are 0.60 m long, with 5 mm diameter holes in their perimeter, spaced 0.50 m apart; to facilitate drainage and prevent soil loading, a synthetic blanket (bidim OP - 20) and a 0.10 m layer of washed gravel n°1 were used as filter material (Figure 4).

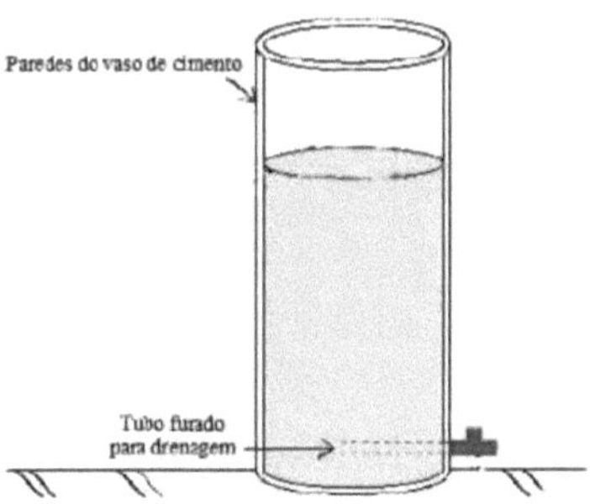

Figura 4 - Diagram of the concrete pipes that formed the experimental plots

To obtain the proposed salinity level, the soils had their electrical conductivity values in the saturation extract raised to the following levels: (S1=2.0: S2=4.0: S3=6.0: S4=8.0 and S5=10.0 dS m^{-1}), by adding saline nutrient solutions dissolved in water from the ESALQ supply network. The amount of salt used to prepare each solution was calculated based on the artificial salinization curve of the respective soil (DIAS, 2001).

The composition of the solution used for salinization was the same as that used in the salinization curve, based on the proportion of Hogland's nutrient solution described by Furlani (1999) (Table 2).

The saline solution was applied using a drip irrigation system with a flow rate of 8 L h^{-1} . The system was made up of self-compensating flow emitters. In order to achieve uniform distribution of the solution on the soil surface, each dripper was connected to a two-discharge divider. At the end of the applications, the levels of electrical conductivity in the soil solution were measured by analyzing the solutions extracted from the soil 48 hours after salinization. Once this was done, the columns were covered with polyethylene plastic to prevent evaporation, remaining so until the soil reached a humidity close to field capacity.

Soil was collected from the pipes for physical and hydraulic analysis. The soil's physical properties were analyzed using the methodology described by Embrapa (2009). The granulometric fractions were determined from the proportions of coarse sand, fine sand, silt and clay, using the pipette method, and expressed as a percentage; the density of soil particles (Dp) was determined using the volumetric flask method; and the density of the soil (Ds) was determined using the measuring cylinder method for deformed samples, both expressed as g cm^{-3} (Table 3).

Table 3 - Physico-hydric characterization of the soil used in the experiment: moisture capacity field (CC) and permanent wilting point (PMP), available water (AD), soil density (Ds), soil particle density (Dp), macroporosity (MP) and total soil porosity (PT).

CC	PMP	AD	DS	Dp	MP	EN	Granulometry			Class
							Sand	Silt	Clay	Textural

cm^3 cm^{-3}			g cm^{-3}		%					
0,23	0,09	0,15	1,30	2,64	27	50	73	8	19	Sandy loam

CC and PMP - Moistures determined at tensions of 6.0 and 1500 kPa, respectively.

Before salinization, soil was collected at depths of 0.00-0.20; 0.20-0.40; and 0.40-0.60 m for the purpose of chemically characterizing the soluble ions (Table 4); to prepare the solutions, a saturation paste was used at saturation humidity, from which the solution was extracted using suction, according to the methodology proposed by Richards (1954).

Table 4 - Chemical characterization of the soluble ions in the soil before the salinization process

Depth	pH (Paste)	EC	In^+	K^+
m		dS m^{-1}	mg L^{-1} _	
0,00-0,20	8,27	1,04	82,50	46,40
0,20-0,40	7,87	1,29	106,60	35,70
0,40-0,60	7,97	1,02	61,20	24,00

EC: Electrical conductivity of the saturation extract

For the chemical characterization of the soil, the phosphorus (P) and potassium (K) contents were determined using the cation exchange resin method and expressed in (mg dm^{-3}). The levels of exchangeable calcium (Ca^{2+}) and magnesium (Mg^{2+}) and potential acidity (H +Al) were extracted from the soil with a 1 mol L ammonium acetate solution^{-1} and determined by flame emission photometry (EMBRAPA, 1997) and expressed in (cmolc dm^{-3}). From these values, the Cation Exchange Capacity (CEC at pH 7.00) and the sum of bases (SB) were determined, expressed in (cmolc dm^{-3}); base saturation (V) was expressed in %.

Table 5 - Chemical characterization of the soil used in the experiment

pH	MO	P	S	K	Ca	Mg	Al	H +Al	SB	CTC	V	M
	g dm^{-3}	---mg dm $^{-3}$			mmolc dm $^{-3}$----						------% -----	
		-		---							-	
6,3	11	174	16	3,4	5,1	27	< 1	15	81,4	96,4	84	1

3.4 Calculation and application of leachate washing slates

The process of recovering the salinized soils consisted of applying sheets of water to leach out the salts and reduce the electrical conductivity, based on an empirical recommendation on the dynamics of solutes. This recommendation states that two pore volumes should be used to calculate the recovery washing slurry for saline soils. As a first step, an initial blade was calculated to bring the soil in the columns up to field capacity, and only then were the recovery washing blades applied. In both systems, a volume of

176 L of water per pipe, and in the intermittent system this volume was divided into two applications, with an interval of six days per application. Right after the applications, black polyethylene sheeting was placed over the pipes to prevent the water from evaporating.

The water used in the applications came from the school's campus water supply, since the test site already had a pipe with access to this system. The EC of this water was considered negligible.

3.5 Measuring soil salinity after washing

Changes in the soil's chemical characteristics, in response to the application of the washing slurries, were assessed by monitoring the electrical conductivity and the concentration of ions remaining in the soil solution of each column. To do this, the soil solution was obtained using extractors and then corrected to saturation humidity.

To this end, three ceramic capsule extractors were installed in the soil of each of the columns (Figure 5). Thus, 180 units of this device were used, which basically consists of a PVC tube with a porous capsule at the bottom. The capsules were positioned at three depths (20, 40 and 60 cm) in the soil profile.

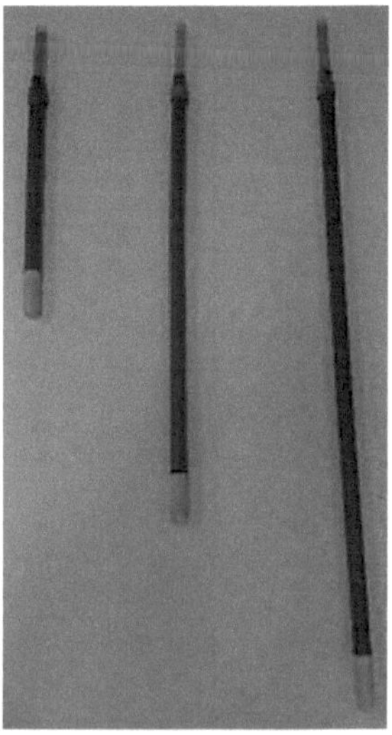

Figure 5 - Details of the ceramic capsule extractors installed in the pipes

To obtain the soil solution, a pressure of 80 KPa was applied to the chamber of each extractor, using a vacuum pump, 48 hours after the end of the application of the blades. After 24 hours, necessary for the redistribution of the solution that was in equilibrium between the capsule and the soil to occur, the solutions were collected from the extractors using a hospital syringe; these were stored in acrylic tubes and then taken for analysis to a laboratory in the Department of Biosystems Engineering.

In the laboratory, the electrical conductivity of each sample was determined using a bench conductivity meter, as well as the concentration of the main ions. The electrical conductivity and ion concentration values, determined using the solution extracted by the porous capsules, were corrected for saturation humidity using eq. (08):

Once the salt concentration (C) needed to prepare the solution had been estimated using the equation based on the salinity curve, the values were corrected for the saturation humidity. Thus, the corrected concentration, which associates the EC read from the extractors with the EC of the saturation stratum, was obtained by applying the expression shown in eq. (08).

$$C_{CORRIGIDA} = \frac{C_{cp} \, . \, U_{cp}}{U_s}$$

(08)

in which:

$C_{CORRIGIDA}$ - Electrical conductivity or concentration of ions in the saturation extract, estimated from the values determined in the soil solution, dS m^{-1} or mmoc L ;$^{-1}$

C_c P - Electrical conductivity or concentration of ions in the soil solution, obtained with a porous capsule extractor, dS m^{-1} or $mmol_c$ L ;$^{-1}$

Ucp - Soil moisture during application of tension to the extractor capsule, g g^{-1} ; e

US - Soil saturation moisture (saturation paste) g g .$^{-1}$

3.6 Evaluation of the performance of empirical equations

Once the tests in the field experiment had been completed, the empirical equations mentioned above were used to estimate the EC of the soil solution after the application of the washings. For comparison purposes, the same leaching rates applied in the real tests were used in the calculations; the initial chemical conditions were also the same as those measured in the soils stored in the percolation columns.

In this way, the EC values were calculated using the equations (01, 02, 03, 04 and 05); to make the calculations easier, the variables of interest in each of the equations (09, 10, 11, 12 and 13) were isolated, respectively. It is important to note that some equations present the results in electrical conductivity while others present them in ion concentration. In this case, the results in EC were converted into concentration using the correlation between C and ECs obtained in the initial phase of the experiment.

$$CF_f = CE_i . 10^{-\left(\frac{L}{\alpha}\right)}$$

(09)

in which:

CEf - Final electrical conductivity in the saturation extract, dS m^ ;^{1}

CEi - Initial electrical conductivity in the saturation extract, dS m ;$^{-1}$

L - Amount of water to be applied to the soil to leach the salts, mm; and

a - Coefficient depending on chloride content and soil texture.

$$CE_f = CE_i \ . \ 10^{-\left(\frac{L}{2,025 \ . \ (P) \ . \left(\frac{70-CE_r}{CE_i-CE_r}\right)}\right)} \quad (10)$$

in which:

CEf - Final electrical conductivity in the saturation extract, dS m ;$^{-1}$

CEi - Initial electrical conductivity in the saturation extract dS m ;$^{-1}$

L - Amount of water to be applied to the soil to leach the salts, mm;

p - Depth of soil to be reclaimed, cm;

CEr - Electrical conductivity of the wash water, dS cm^{-1} ; and CEi - Initial electrical conductivity, dS m .$^{-1}$

$$C = 0,8 \ . \ C_o \ . \ \theta \ . \left(\frac{D_s}{D_e}\right) \quad (11)$$

in which:

C - concentration of salts in the soil solution after recovery, mg L ;$^{-1}$

Co - concentration of salts in the soil solution before recovery, mg L ;$^{-1}$

θ - volumetric soil moisture, cm^3 cm ;$^{-3}$

Ds - depth of soil to be reclaimed, cm; and

De - làmina of water to be applied to the soil to wash off the salts, cm.

$$C = C_o \ . \ K \ . \left(\frac{D_s}{D_e}\right) \quad (12)$$

in which:

C - Concentration of salts in the soil solution after recovery, mg L^ ;[1]

C_o - Concentration of salts in the soil solution before recovery, mg L^{-1} ; K - Constant that varies with the type of soil and the method of water application;

DS - Depth of soil to be reclaimed, cm; and

D_e - Amount of water to be applied to the soil to wash off the salts, cm.

$$CEf = \frac{\left(\frac{La}{Ls} - 0{,}15\right).CEi}{5} \qquad (13)$$

in which:

CEf - Concentration of salts in the soil solution after recovery, mg L ;$^{-1}$

La - Height of water to be leached, cm;

Ls - Thickness of soil to be recovered, cm; and

CEi - Concentration of salts in the soil solution before recovery, mg L.

The index of agreement (CAMARGO & SENTELHAS 1998) and the root mean square error were used to measure the performance of the equations used in this research.

The parameters evaluated were also statistically analyzed using analysis of variance and the Tukey mean test at 5 and 1% probability levels. Data analysis was carried out in the R language and environment, 2.15.1 (R DEVOLOPEDMENT CORE TEAM 2013).

The performance of the equations was evaluated by comparing the final EC obtained in the calculations using these equations with the experimental results measured in the test, in the pipes installed in the experimental area.

4 RESULTS AND DISCUSSION

4.1 Classical statistics for the performance of empirical equations

According to the analysis of variance, there was a significant interaction ($p< 0.01$) between the factors recovery system and salinity levels for the data obtained, only for the pore volume wash "Experiment", and an isolated effect of salinity for both the experiment and all the equations studied; for the recovery systems there was no significance in the equations (Table 6). These results show that the reclamation systems have no effect on the equations studied, as most of them do not go as far as to indicate which system will be used to reclaim the soil.

Table 6 - Summary of the analysis of variance (F test) and statistical significance for the values of the experiments and for the variables estimated EC and measured EC in the recovery systems

Statistics (F)								
FV	GL	Experiment	Volubuyev	Lamb	Jury	Hoffman	Reeve
System	1	185,94**	$0,58^{ns}$	$2,66^{ns}$	$0,58^{ns}$	$0,58^{ns}$	$0,58^{ns}$
Salinity	4	67,69**	2078,31**	701,91**	2078,77**	2077,25**	2078.01**
Sis x Sali	4	12,82**	$2,16^{ns}$	$1,65^{ns}$	$2,16^{ns}$	$2,16^{ns}$	$2,16^{ns}$
CV (%)	-	18,317	3,897	1,879	3,896	3,898	3,897

**significant at 5% probability and^{ns} not significant

A significant effect was observed for the interaction between recovery systems and salinity levels for the values obtained in the experiment, only for the "Experiment" pore volume wash (Table 06). These data are consistent with those obtained by Barros (2005) and De Jong (1982).

Working with saline soils, these authors found that 50% of the dissolved salts were leached out when a leaching slurry equal to 1.0 pore volume was applied. Similar results were obtained by Silveira (2000); using gypsum dissolved in irrigation water to recover saline and saline-sodic soils in Custódia, PE, he observed that the first leachate collected from the soil columns showed the highest EC values, decreasing successively until it equaled the EC of the water used for irrigation.

If the system-salinity interaction is broken down, it can be seen that for both types of systems there was an increase in the salinity level as the salinity increased; it was also found that there was variation according to the system used. In general, the lowest values were obtained in the intermittent system, which is based on the gradual washing away of salts, thus allowing more of the salts that are retained in the soil to be washed away; while continuous management, despite the short application time, is not as efficient at removing these salts from the soil.

The movement of salts in the soil solution takes place through the flow of ions in the soil, which occurs through two processes: diffusion and mass transport. According to Reichardt (1996) and Libardi (2012), it is common for the processes of mass flow and diffusion to occur simultaneously, when water moves through infiltration into the soil and through the difference in concentration caused by the addition of salts after a fertigation event.

The highest average salinity values were found in the continuous system (Table 7). Comparing the systems within each salinity level, it can be seen that for all salinity levels the systems do not differ statistically. In general, for the systems studied, there was a gradual concentration effect at the salinity levels (Table 7).

This salinity interaction had an influence on the results, possibly due to the granulometry of the soil, which is an intermediate-textured soil. The increase observed is in line with what was expected and proposed, for example, by van der Sloot et al. (1997), who studied the influence of particle size in leaching tests. These authors found that larger specific surface areas, per unit mass or volume, can allow for rapid dissolution on the surface.

Table 7 - Average salinity values in the experiment in the recovery systems SALINITY LEVELS (dS m)$^{-1}$

SiSTEMS	2,00	4,00	6,00	8,00	10,00
intermittent	0.562 Ce	0.646 Cd	0.712 BCc	0.986BAb	1,438 Aa
Continued	0.876 Ce	1.181 Cd	1,641 Bc	1.836 Bb	2.867 Aa

Values followed by the same lower-case letter in the column and upper-case letter in the row do not differ by the Tukey test at 5% probability.

4.2 Performance of equations for predicting soil salinity in the intermittent system

Figure 06 shows the five dispersions of the points representing the measured and simulated electrical conductivity values around the 1:1 line, as well as the regression equation for the five washing equations in the intermittent recovery system.

Figure 06 shows an adjustment of R^2 =0.9281 for A; 0.9081 for C; 0.9481 for D; 0.9088 for E and 0.8387 for B, between the electrical conductivity values measured in the saturation extract (estimated from the electrical conductivity of the solution) and the calculated values. The measured values were obtained with porous capsule extractors, corrected for humidity, in the intermittent recovery system; while the estimated values were calculated using the Volobuyev - A; Cordeiro - B; Jury - C; Hoffman - D and Reeve - E equations. According to Figure 6A, using the Volobuyev recovery equation, the estimate of electrical conductivity is close to that obtained with the measured values, with the trend line almost overlapping the 1;1 line. These results show that the Volubuyev equation is able to estimate with good accuracy, indicating that the method is reliable for estimating EC in the intermittent recovery system. In comparison, the estimated values in Figures 6B, 6C, 6D and 6E performed less well. Figures 6B, 6C and 6E correspond to the equations that showed the greatest overestimation of electrical conductivity values for this recovery method, while the values found by the Hoffman equation underestimated the measured electrical conductivity values of the extract.

In general, the equations used tended to overestimate the measured values, especially for the highest values, such as the Jury equation; the Cordeiro equation showed the greatest discrepancies. This behavior is possibly due to the fact that this equation is highly sensitive to the initial levels of soil salinity.

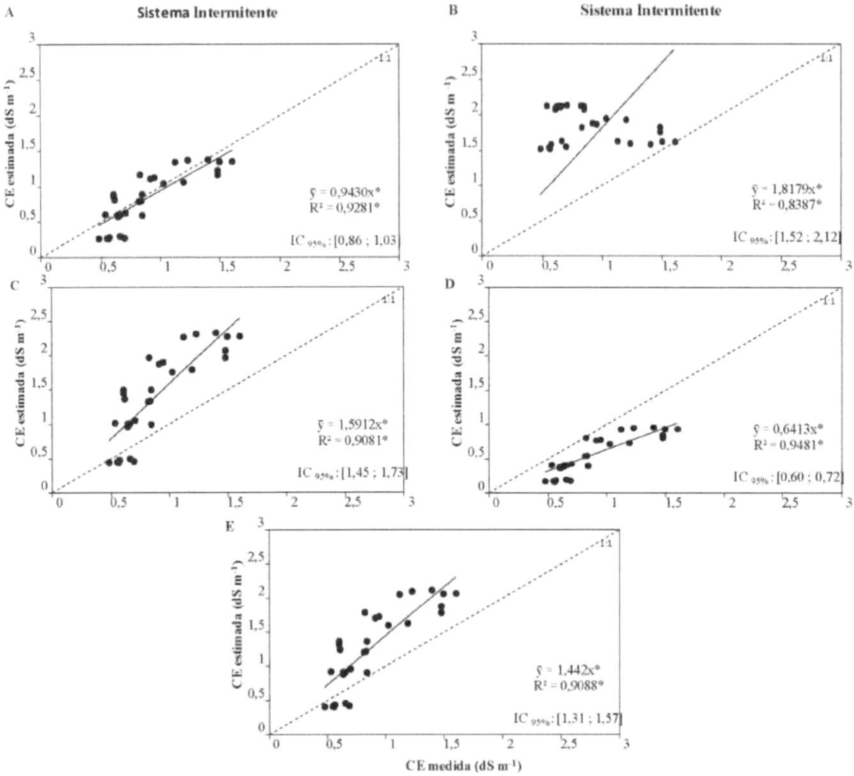

* **Significantly> 1 at 5% Probability level,** by "t" test

Figure 6 - Relationship between the values of electrical conductivity of the saturation extract estimated with the equations and the values measured from the electrical conductivity of the soil solution obtained with porous capsule extractors, for intermittent recovery, (A - Volubuyev; B - Cordeiro; C - Jury; D - Hoffman and E - Reeve)

The values of the coefficients of determination, Pearson's linear correlation, Willmott's concordance index, Camargo Sentelhas' performance index, the standard error of the estimate and the root mean square can be seen in Table 08.

Table 8 - Values of the coefficients of determination (R^2 a), Pearson's linear correlation (r), Willmott's concordance index (w), Camargo Sentelhas' performance index (c), the standard error of the estimate (EEP) and the root mean square of the error (RQME) obtained in the correlations between the estimated and measured EC values in the intermittent recovery system.

Equations	P-Value	R^2	r	w	c	EEP	RQME

Volobuvev	<0,001	0,946	0,832	0,999	0,857	0,039	4,874
Lamb	<0,001	0,833	ns	0,868	0,621	0,197	9,396
Jury	<0,001	0,946	0,832	0,945	0,810	0,120	8,224
Hoffman	<0,001	0,946	0,878	0,956	0,819	0,064	3,427
Reeve	<0,001	0,946	0,832	0,965	0,828	0,096	7,452

Table 8 shows equal values for the coefficients of determination for the equations studied; only the Cordeiro equation showed a lower value than the others, with an R^2 of 0.833. For Pearson's linear correlation, only the Cordeiro equation was not significant, while the others were significant at a probability of 5%. This equality occurs because grouping the estimated EC values in relation to the measured ones tends to improve the EC estimates. The methods under study showed high values for Willmott's coefficient, with the values obtained by Volobuyev being the closest to 1, i.e. values closest to perfect agreement. Therefore, the best result obtained by equations for the intermittent recovery system was that of Volobuyev. This equation, as well as having the best performance index according to Camargo and Sentelhas, has the lowest EEP and RQME values, proving that the product of the Iinear dimensions increases the intensity of the linear association between the variables. According to José (2014), who worked with these same parameters, evaluating equations for estimating leaf area in the basil crop, the closer the value of R^2 is to 1.0, the closer the error values are to zero and the greater the accuracy of the adjusted model.

4.3 Performance of the equations for predicting soil salinity in the continuous system

Figure 7 shows the values of the coefficients of determination R^2 = 0.9186 for A; 0.9108 for C; 0.9644 for D; 0.9463 for E and 0.8009 for B. The measured ECs were obtained using porous capsule extractors, corrected for humidity; the estimated values were calculated using the Volobuyev - A; Cordeiro - B; Jury - C; Hoffman - De Reeve - E equations. It can also be seen in Figure 07 that, in general, all the equations studied in this recovery system showed values that underestimated the measured values. The Cordeiro equation was the one that provided the best regression line in relation to the

1:1 line; however, with very discrepant values compared to the others. As these dispersions are characterized by mathematical approximations that evaluate the accuracy, dispersion and distance of the simulated values from the calculated ones, it can be said based on these results that, for this recovery method, the equations are not as efficient when compared to the previous method. Thus, the values in the intermittent system are better described by the empirical equations.

Figure 7B also shows that there is a large dispersion of points around the regression line, which is different from the other equations where the points are closer to the 1:1 line, allowing for good correlation between the measured and calculated values.

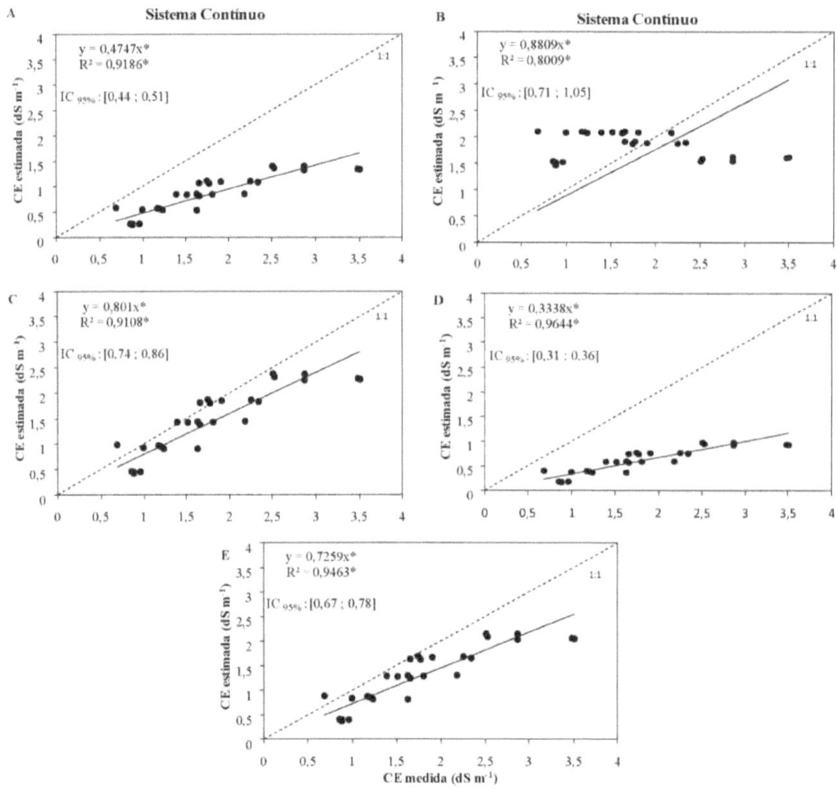

*** Significantly> 1 at 5% Probability level, by "t" test**

Figure 7 - Relationship between the values of electrical conductivity of the saturation extract estimated by the equations and the values measured from the electrical conductivity of the soil solution obtained with porous capsule extractors, for continuous recovery, (A - Volubuyev; B - Cordeiro; C - Jury; D - Hoffman and E - Reeve)

The values of the coefficient of determination, Pearson's linear correlation coefficient, Willmott's concordance index, the Camargo & Sentelhas performance index, the standard error of the estimate and the root mean square for the continuous recovery system can be seen in Table 09.

Table 9 - Values of the coefficients of determination (R^2), Pearson's linear correlation (r), Willmott's concordance index (w), Camargo Sentelhas performance index (c), standard error of estimate (SEE) and root mean square error (RMSE) obtained in the correlations between estimated EC values and measured EC values in the intermittent recovery system.

Equations	P-Value	$R2_a$	r	w	c	EEP	RQME
Volobuvev	<0,001	0,963	0,898	0,870	0,799	0,173	4,913
Lamb	<0,001	0,794	ns	0,999	0,702	0,157	9,116
Jury	<0,001	0,963	0,897	0,989	0,908	0,168	8,289
Hoffman	<0,001	0,963	0,898	0,735	0,675	1,209	3,454
Reeve	<0,001	0,963	0,898	0,977	0,987	0,094	7,512

Table 9 shows equal values for the coefficients of determination for the equations studied; only the Cordeiro equation showed a lower value than the others, with an R^2 of 0.794. For Pearson's linear correlation, only the Cordeiro equation was not significant, while all the others were significant at a probability of 5%. The methods under study: Cordeiro, Jury and Reeve were the ones that showed the best values for the Willmott coefficient, with the values obtained by Cordeiro being the method that came closest to 1, a value that characterizes perfect agreement. The PDE values were lowest in the Reeve equation (0.094), followed by Cordeiro (0.157); Jury (0.168) and Volobuyev (0.173), with the Hoffman equation having the highest value, at 1.209. For the RQME, Hoffman's equation also provided the lowest value.

The results found in this research disagree with the values obtained by Franco (2013). This author, working with four of the five equations used in this research and with the same soil and two levels of salinity, suggested that the best equation would be that of Cordeiro, for calculating the leaching rate of salts in salinized soils. The only difference is that Franco (2013) salinized the soils with sodium chloride.

Franco (2013) also points out that the values obtained by the equations differ from some studies in the literature, and that these are quite variable, which may be related to the diversity of soils and the conditions in which the studies were carried out. This shows that the results obtained by the recovery equations possibly depend on physico-chemical factors related to the soils or inherent in the methodology itself, and that these prevent consistent results from being obtained under different experimental conditions.

It seems that future research will need to consider the formulation of the equations and the physical-chemical aspects of the soil, and not just the initial and final electrical conductivity values of the soil and the water used for washing.

5 CONCLUSIONS

Given the conditions in which the experiment was conducted, it can be concluded that:

Technically, it is possible to use salinized soil reclamation washing as an irrigation management strategy, and it is very satisfactory;

Among the equations evaluated, Volobuyev's was the one that showed the best performance in recovering salinized soil, providing answers that were more consistent with the results obtained experimentally; and

In general, the equations tested were more efficient in the intermittent recovery system; in the continuous system, the calculated electrical conductivity values after washing were all underestimated when compared to those measured experimentally.

REFERENCES

AYERS, R.S.; WESTCOT, D.W. **Water quality in agriculture.** Campina Grande: UFPB, 1991. 218p.

BARROS, M. F. C.; SANTOS, P. M.; SILVA, A.J. Recuperaçao de solos afetados por Recuperaçao de solos afetados por salais usando agua de qualidade inferior salais usando agua de qualidade inferior. **Revista Brasileira de Engenharia Agricola e Ambiental**, Campina Grande, PB. v.9, (Suplemento), p.310-313, 2005.

BECKER, B.K. Geopolitics of Amazonia. **Revista Estudos Avançados,** Sao Paulo, v. 19, n. 53, p. 71-86, jan./abr. 2005.

BERNARDO, S. **Manual de irrigaçâo**. 6. ed. Viçosa: UFV, 1996. 596p.

BERNARDO, S; SOARES, A.A; MANTOVANI E.C. **Manual de irrigaçâo**. 7. ed. current and expanded. Viçosa: Editora UFV, 2005. 611p.

BISHOP, A.A.; WALKER, W.R.; ALLEN, N.L; POOLE, G.J. Furrow advances rates under surge flow systems. **Journal of the Irrigation and Drainage Division**, ASCE, v.107, n.3, p.257-264, 1981.

BOT, A.; NACHTERGAELE, F.; YOUNG, A. **Land resource potential and constraints at regional and country levels.** Rome: FAO, 2000.180p. (FAO. World Soil Resources Report, 90).

BLANCO, F.F. ; FOLEGATTI, M.V. Recovery of a salinized soil after cultivation in a protected environment. **Revista Brasileira Engenharia Agricola**, Campina Grande, PB. v.5, n.1, p. 76-80, 2001

CAMARGO, A.P.; SENTELHAS, P.C. Evaluation of the performance of different methods for estimating potential evapotranspiration in the state of Sao Paulo.

Revista Brasileira de Agrometeorologia, Santa Maria, v. 5, n. 1, p. 89-97, 1997.

CORDEIRO, G.G. **Salinity in irrigated agriculture (basic and practical concepts).**

Petrolina, PE: Embrapa Semiàrido, 2001. 38p.

COOLIDGE, P.S.; WALKER, W.R.; BISHOP, A.A. Advance and run-off surge flow furrow irrigation. **Journal of the Irrigation and Drainage Division**, ASCE, v.108, n.1, p.35-42, 1982.

DAKER, A. **Agua na agricultur:** manual de hidrâulica agricola. 7.ed. Rio de Janeiro, RJ: Freitas Bastos, 1988. 543p.

DE JONG, E. Reclamation of soils contaminated by sodium chlorid. **Canadian Journal Soil Science**, Ottawa, v.62, n.2, p.351-364, 1982.

DIAS, N.S. **Prevention, management and recovery of soils affected by salts**.

DIAS, N. da SILVA X GHEYI, H.R.; X ;DUARTE,S.N. Piracicaba: ESALQ/USP/LER, 2003.118p. (Série Didâtica, 13).

DIAS, E.G.C.S. **Environmental Impact Assessment of mining projects in the State of São Paulo: the follow-up stage**. 2001. 206p. Thesis (Doctorate in Mineral Engineering) - Polytechnic School of the University of Sao Paulo, Sao Paulo, 2001.

EMBRAPA. National Soil Research Center. **Brazilian Soil Classification System.** 2. ed. Rio de Janeiro : Embrapa Solos, 2006. 306p.

EMBRAPA. **Manual of chemical analysis of soils, plants and fertilizers**. Embrapa Informaçao Tecnológica. 2. ed. revised and expanded. 2009. 167p.

FRANCO, E.M. **Efficiencies of empirical equations used to determine the leaching rate of salts and modeling of sodium distribution in soil**. 2013. 54p. Thesis (Doctorate in Irrigation and Drainage) - "Luiz de Queiroz" School of Agriculture, University of São Paulo, Piracicaba, 2013.

FAO. **Water in agriculture**: opportunity untapped. Rome: Food and Agriculture Organization of the United Nations, 2006.203p.

FURLANI, P.R.; SILVEIRA, L.C.P.; BOLONHEZI, D.; FAQUIN, V. **Hydroponic**

cultivation of plants. Campinas: IAC, 1999. 52p. (Technical bulletin, 180).

GHEYI, H.R. Salinity problems in irrigated agriculture. In: OLIVEIRA, T.S.; ASSIS, R.N.; ROMERO, R.E.; SILVA, J.R.C. (Ed.). **Agriculture, sustainability and the semi-arid** region. Fortaleza: DCS/UFC, 2000. p. 329-346.

HOFFMAN, G.J.; RHOADES, J.D.; LETEY, J.; SHENG, F. Salinity management. In: HOFFMAN, G.J.; HOWELL, T.A.; SOLOMON, K.H. (Ed.) **Management of farm irrigation systems**. St. Joseph: ASAE, Pamela De-Vore-Hansen, 1992. ch.18 p.667-715, (ASAE Monograph.).

HOFFMAN, G.J. Alleviating salinity stress. In: ARKIN, G.F.; TAYLOR, H.M. (Ed.). **Modifying the cRoot environment to Reduce Crosp stress**. St. Joseph, Ml: American society of Agricultural Engineers, 1980. chap.9 p. 305-343.

HOLANDA, J.S.; AMORIM, J.R.A.; FERREIRA NETO, M.; HOLANDA, A.C. Quality of water for irrigation. In: GHEYI, H.R.; DIAS, N.S.;

LACERDA, C.F. **Management of salinity in agriculture:** Basic and applied studies**.** Fortaleza: National Institute of Science and Technology in Salinity, 2010. p. 43-61.

JOSÈ, J.V. **Potassium fertilization and irrigation rates on the production of biomass and essential oil in *basil* (*Ocimum basilicum* L.).** 2014. 115p. Thesis (Doctorate in Irrigation and Drainage) - Escola Superior de Agriculturas "Luiz de Queiroz" , Universidade de Sâo Paulo, Piracicaba, 2014.

JURY, W.A.; JARRELL, W.M.; DEVITT, D. Reclamation of saline-sodic soils by leaching. **Soil Science Society of America Journal**, Madison, v. 43, p. 11001106, 1979.

KEMPER, W.D.; TROUT, T.J.; HUMPHERYS, A.S.; BULLOCK, M.S. Mechanisms by Which Surge Irrigation Reduces Furrow Infiltration Rates in a Silty Loam Soil. **Transaction of the ASAE,** city, v.31, n.3, p.821-829, 1988.

LIBARDI, P.L. **Dynamics of water in the soil**. Sâo Paulo: Edusp, 2012. 344p.

LIMA, V.L.A. **Effects of irrigation water quality and leaching fraction on the bean crop (Phaseolus vulgaris L.) under drainage lysimeter conditions.** 1998. 87p. Thesis (Doctorate in Agricultural Engineering) - Federal University of Viçosa, Viçosa, 1998.

MACHADO, S.L.O. MARCHEZAN, E. ; VILLA, S.C.C. ; CAMARGO, E.R. Os recursos hidricos e a lavoura arrozeira e recursos hidricos. **Ciência & Ambiente**, Santa Maria, n. 27, p. 97-106, 2003.

MONTENEGRO, A.A.A.; MONTENEGRO, S.M.G. Sustainable use of alluvial aquifers in the semi-arid region. In: CABRAL, J.S.P.;

FERREIRA, J.P.C.L.; MONTENEGRO, S.M.G.L.; COSTA,W.D. **Agua subterrànea**: Aqüiferos costeiros e aluvioes, vulnerabilidade e aproveitamento. Special topics in water resources. Recife, Federal University of Pernambuco, 2004. v.4. 447p.

OLIVEIRA, M. Genesis, classification and extent of soils affected by salts. In: GHEYI, H.R. QUEIROZ, J.E.; MEDEIROS, J.F. **Manejo e controle da**

salinity in irrigated agriculture. Campina Grande: UFPB/SBEA, 1997. chap.1, p.1-35.

Genesis, classification and extent of soils affected by salts. In: SYMPOSIUM "MANAGEMENT AND CONTROL OF SALINITY IN IRRIGATED AGRICULTURE", 1997, Campina Grande, PB. **Anais...**Campina Grande: UFPB , 1997. p. 3-35.

PALACIOS, O.V.; ACEVES, E.N. **Instructivo para el muestreo registro de dados e interpretacion de calidad del agua para Riego agricola**. Chapingo: College of Postgraduates, National School of Agriculture, 1969. 498p. (Riego y Drenaje Branch, 15).

PEREIRA, J.R. Soils affected by salts. In: CAVALCANTI, F.J.A. (Coord).

Fertilization recommendations for the State of Pernambuco: 2ª approximation. 2 ed. Recife: IPA, 1998. p. 76-82.

PIZARRO, F. **Drenaje agricola y recuperacion de suelo salinos**. Madrid: Agricola Espanola, 1978. 521p.

PIZARRO, R. **Rasgos y actitudes del profesor efectivo.** Thesis for the Degree of Magister in Educational Sciences Pontificia. Universidad de Chile, 1996, Chile.

PRISCO, J.T. Some aspects of the physiology of salt stress. **Revista Brasileira de Botânica**, Sao Paulo, v. 3, p. 85-94, 1987.

QADIR, M.; QURESHI, R.H.;AHMAD,N. Reclamation of a saline-sodic soil by gypsum and Leptochloa fusca. **Geoderma,** Amsterdam, v.72, p.207-217, 1996.

R DEVELOPMENT CORE TEAM R: language and environment of statistical computing. Vienna, 2013.

REEVE, R.C. **The relation of salinity to irrigation and drainage requeriments**. Int. Comm. Irrig. Drainage Thirdc Congr. Q. 10, 1957. p.175187.

(US. Saline Lab., riverside Calif).

REICHARDT, K. Dynamics **of matter and energy in ecosystems**. 2.ed. Piracicaba: Ed. ESALQ, 1996. 513p.

RHOADES, D.J. Drainage for salinity control. In: SCHNFGAARDE, J.VAN. (Ed.). **Drainage for agriculture**. Madison, Wisconsin: American Society of Agronomy, 1974. p.433-460. (Agronomy,17).

RHOADES, J.D.; KANDIAH, A.; MASHALI, A.M. **The use saline water for crop production.** Rome: FAO, 1992. 133p. (FAO Irrigation and Drainage Paper, 48).

RHOADES, J.D.; LOVEDAY, J. Salinity in irrigated agriculture. In: STEWART, D.R.; NIELSEN, D.R. (Ed.) **Irrigation of agricultural crops**. Madison: ASA, CSSA, SSSA, 1990. p. 1089-1142. (Agronomy, 30).

RIBEIRO, M.R.; BARROS, M.F.C.; FREIRE, M.B.G.S. Chemistry of saline and sodic soils. In: MELO, V.F.; ALLEONI, L.R.F. (Ed.). **Soil chemistry and mineralogy.** Part II - Applications. Viçosa: Brazilian Society of Soil Science, 2009. p. 449-484.

RIBEIRO, M.R.; FREIRE, F.J.; MONTENEGRO, A.A.A. Halomorphic soils in Brazil: occurrence, genesis, classification, use and sustainable management. In: CURI, N.; MARQUES, J.J.; GUILHERME, L.R.G.; LIMA, J.M.; LOPES, A.S;

ALVAREZ, V.H. (Ed.). **Topics in Soil Science**. Viçosa: Brazilian Society of Soil Science, v.3, 2003. p.165-208.

RICHARDS, L.A. **Diagnosis and improvement of saline and alkali soils.** Washington: United States Department of Agriculture, 1954. 160p. (Agriculture Handbook, 60).

SANTOS, R.V. dos ; HERNANDEZ, F.F.F.Recovery of soils affected by salts. In: BRAZILIAN CONGRESS OF AGRICULTURAL ENGINEERING - MANAGEMENT AND CONTROL OF SALINITY IN IRRIGATED AGRICULTURE, 26, 1997. Campina Grande . **Proceedings...** Campina Grande: UFPB, 1997. p. 319 - 356.

SOUZA, M.R. de. **Behavior of bean plants (Phaseolus vulgaris L. CV Eriparza) subjected to different levels of irrigation water salinity**.

1995. 94p. Dissertation (Master's Degree in Agricultural Engineering) - Federal University of Lavras, Lavras, 1995.

SILVEIRA, K.R. **Influence of gypsum application in irrigation water on salinity, sodicity and hydraulic conductivity of alluvial soils**.

2000. 73p Dissertation (Master's Degree in Agricultural Engineering) - Federal Rural University of Pernambuco, Department of Rural Technology, Recife, 2000.

STONE, L.F. **Eficiência do uso da água na cultura do arroz irrigado**. Santo Antônio de Goiâs: Embrapa Arroz e Feijao, 2005. 22p.

TAVARES FILHO, A.N. **Levels of the need for gypsum on the physico-chemical characteristics and in the correction of saline-sodic soils in the irrigated perimeter of Ibimirim -PE.** 81p. 2010. Dissertation (Master's Degree in Agricultural Engineering) - Federal Rural University of Pernambuco, Department of Rural Technology, Recife, 2010.

VAN DER SLOOT, H.A.; HEASMAN, L.; QUEVAUVILLER, Ph.; **Harmonization of Leaching/ Extraction Tests.** Amsterdam: Elsevier, 1997. 281p.

WALLENDER, W.W.; RAYEJ, M. Zero-Inertia Surge Model with Wet-Dry Advance. **Transaction of the ASAE,** _Britain, v. 28, n.5, p.1530-1534, 1985.

More Books!

yes

I want morebooks!

Buy your books fast and straightforward online - at one of world's fastest growing online book stores! Environmentally sound due to Print-on-Demand technologies.

Buy your books online at
www.morebooks.shop

Kaufen Sie Ihre Bücher schnell und unkompliziert online – auf einer der am schnellsten wachsenden Buchhandelsplattformen weltweit! Dank Print-On-Demand umwelt- und ressourcenschonend produzi ert.

Bücher schneller online kaufen
www.morebooks.shop

info@omniscriptum.com
www.omniscriptum.com

OMNIScriptum

MIX
Papier aus verantwortungsvollen Quellen
Paper from responsible sources
FSC® C105338

Printed by Books on Demand GmbH, Norderstedt / Germany